W9-CLP-181

YALE
MATHEMATICAL
MONOGRAPHS

2

YALE
UNIVERSITY
PRESS

ELIAS M. STEIN

Analytic Continuation of Group Representations

2.95 N
mc

Yale Mathematical Monographs 2

James K. Whittemore Lectures in Mathematics
given at Yale University

ANALYTIC CONTINUATION OF GROUP REPRESENTATIONS

by Elias M. Stein

New Haven and London, Yale University Press, 1971

Library of Congress catalog card number: 73-151591.

International standard book number: 0-300-01428-7 (paper).

Set in IBM Bold Face One type.
Printed in the United States of America by
The Murray Printing Co., Forge Village, Mass.

Distributed in Great Britain, Europe, and Africa by
Yale University Press, Ltd., London; in Canada by
McGill-Queen's University Press, Montreal; in Mexico
by Centro Interamericano de Libros Académicos,
Mexico City; in Central and South America by Kaiman
& Polon, Inc., New York City; in Australasia by
Australia and New Zealand Book Co., Pty., Ltd.,
Artarmon, New South Wales; in India by UBS Publishers'
Distributors Pvt., Ltd., Delhi; in Japan by John
Weatherhill, Inc., Tokyo.

1287481

Analytic Continuation of Group Representations

Lecture I.   Introduction

1. In any general discussion of the theory of representations and harmonic analysis on Lie groups it is appropriate, at least at one stage, to try to see the problems in a general setting. Let us therefore begin by recalling the general questions which have guided the development in the past and will certainly continue to serve in this role in the foreseeable future. Stated generally these problems are, in the context of any group G, as follows:

Problem A.   What are "all" the irreducible unitary representations of G?

Problem B.   What is the Plancherel formula for G, i.e., what is the analysis of $L^2(G)$, in explicit as possible terms?

Problem C.   (Here we shall be even more vague.) What are the other interesting aspects of harmonic analysis of G?

2. In the case of the Abelian groups, in particular the group of Euclidean n-space, the n-torus, the Bohr group for a.p. functions, etc. the solutions of A and B are classical and very well known. In that context what faces us (and will continue to do so) is problem C. Much the same is true for the compact Lie groups and the nilpotent Lie groups, although in the latter case the solution of problems A and B is more recent and possibly not generally familiar.

Our main concern in these lectures will be with the non-compact semi-simple Lie groups, and so let us specify in general terms what can be said about these problems.

A James K. Whittemore Lecture, November 1967. Originally published in *Advances in Mathematics* 4 (2), April 1970, 172; reprinted with the permission of Academic Press.

1

As far as problem A is concerned, it can be stated that for any such group wide classes of irreducible unitary representation have been constructed, but—and this point seems not to be widely realized—there exists at this time no complete classification of the irreducible unitary representations of any such G, save in the case of certain special groups G of low dimension. We shall have more to say about this matter in the course of these lectures.

As to the problem of the Plancherel formula matters stand as follows: When G is a *complex* semi-simple group, the Plancherel formula is known explicitly as a result of the work of Gelfand and Neumark and Harish-Chandra. As to the general semi-simple case there is as yet no explicit Plancheral formula, but we are probably not far from that goal, thanks to the heroic labors of Harish-Chandra. We shall have nothing to add to this problem here, although we shall find a use for the explicit Plancherel formula.

As to the other problems of harmonic analysis on semi-simple groups it is much harder to give a complete survey of this area. Suffice it to say that much progress has been made in the study of G/Γ, where Γ is an appropriate discrete subgroup. The results center about automorphic forms, Eisenstein series, trace formula of Selberg, etc.

It is not this area of application that will concern us here, but rather other types of questions of the harmonic analysis of G that we will deal with.

3. We come now to the subject matter proper of these lectures. A few remarks of a personal nature may not be out of place here. The results that will be presented here are the fruits of a collaboration with my colleague R. Kunze begun nearly ten years ago and still in progress.[1] As the listener may have already surmised the main results obtained do not fit neatly into the simple pattern described above. It is my belief that, nevertheless, the results presented here have a definite interest and are therefore worthy of your serious consideration.

We begin by some notation. Let G be a semi-simple group of non-compact type with finite center; let K be a maximal compact subgroup, and $G = KAN$, a corresponding Iwasawa decomposition.[2] If M is the

---

[1] The published results are [1], [2], and [4]. The paper [3] appeared only in preprint form, and most of its contents was subsumed in a later publication. The paper [6] will appear soon; other papers are in preparation.

[2] For further details on the general properties used here see Lecture III, Sections 1 and 2.

centralizer of A in K we consider the subgroup $B = MAN$ which plays a fundamental role in all that follows. For any irreducible unitary finite dimensional representation λ of B, we form the representation $g \to T_g^\lambda$ which is the representation of G induced by λ. The family of representations $\{T^\lambda\}$ so obtained are the "non-degenerate principal series of representation."

Their importance is due to the following facts:

(a) They are in general irreducible.

(b) When G is a *complex* semi-simple group, these representations suffice for the Plancherel formula of the group, i.e., the $L^2(G)$ analysis.

(c) For the case of G a general real group, other representations must also be brought into play for the Plancherel formula, but the other series unlike the principal series have not as yet been constructed in global form, except in special cases.

A closer look at these representations show that they depend in a natural way on certain discrete parameters (the rank of M) and a certain number of continuous parameters (the dimension of A, which is also called the rank of the symmetric space defined by G; we shall call this number r).

More precisely A is the product of r copies of the multiplicative group of positive numbers. That is we can represent A as $(\delta_1, \delta_2, ..., \delta_r)$, with $\delta_i > 0$. Now any unitary irreducible representation λ of B is uniquely determined as a product of representations of M and A, respectively, both of which must again be unitary and irreducible. The unitary irreducible representatives of the compact group M are determined by the discrete parameters mentioned, but the unitary irreducible representatives of A are of course the mappings $(\delta_1, \delta_2 \cdots \delta_r) \to \delta_1^{it_1} \cdots \delta_r^{it_r}$ where $t_1, t_2, ..., t_r$ are r arbitrary real numbers.

It is now clearly indicated (if not clearly indicated, at least no analyst can resist the temptation!) that we should want to continue in the parameters $it_1, ..., it_r$ to more general complex-values.

On a certain formal level, once the induced representatives T^λ have been given an explicit form, the analytic continuation is obvious. But lest the listener be misled he must be warned that this naive analytic continuation is irrelevant, because, unlike the context of ordinary functions, in the setting of representations analytic continuations are highly non-unique! The reason for this is that we are in fact continuing representatives from equivalence classes of representations and of course two representations may be equivalent without being identical.

Let us write $\lambda = \lambda(m, s) = \lambda_M \cdot \lambda_s$, where λ_M is the component of the representation λ belonging to the compact group M; this representation is parameterized by the discrete parameter m. λ_s is the component of the representation belonging to A; i.e., $\lambda_s(\delta) = \delta_1^{s_1} \delta_2^{s_2} \cdots \delta_r^{s_r}$. Another important element must still be defined, the Weyl group. Let M' be the normalizer of A in K. Then M'/M is a finite group and every element $p \in M'/M$ induces a natural action $\lambda \to p(\lambda)$, with $p(\lambda) = \lambda(p^{-1}m, p^{-1}s)$. We define also the complex contragredient character λ' by

$$\lambda'(m, s) = \lambda(m, -\bar{s}) \qquad (\bar{s} = \text{complex conjugate of } s).$$

We recall the definition of a *tube* in the complex r-space, with basis B, $s_j = \sigma_j + it_j$, $j = 1,..., r$. B is an open convex space in the r dimensions of the $\sigma_1,..., \sigma_r$, and $T_B = $ tube with basis $B_B = \{s \mid \{\sigma_j\} \in B\}$. The main results to be described are consequence of the following theorem:

THEOREM 1. *Let G be a group of the kind detailed below. Then there exists a tube T_B described above whose basis contains the origin (in its interior), and a family of bounded operators in a Hilbert space $\mathscr{H}$, $\{R_g^{\lambda(m,s)}\}$ with the following properties:*

(1) *for each fixed m, and $s \in T_B$, $g \to R_g^{\lambda(m,s)}$ is a cont. rep. of G.*

(2) *For each fixed g, $mR^{\lambda(m,s)}$ is analytic (as a function of s) in the tube T_B.*

(3) $\sup_{g \in G} \| R_g^{\lambda(m,s)} \|$ *is finite, i.e., the representations are uniformly bounded. Moreover, $\sup_{g \in G} \| R_g^{\lambda(m,s)} \|$ is at most of polynomial growth in the parameters m and $it_1,..., t_r$.*

(4) *When $R(s_1) = R(s_2) \cdots = R(s_r) = 0$, the representations $R_g^{\lambda(m,s)}$ are unitarily equivalent to the principal series $T_g^{\lambda(m,s)}$.*

(5) *For any element p of the Weyl group*

$$R_g^{p\lambda} = R_g^{\lambda}.$$

(6) *For any $\lambda(m, s)$, $s \in T_B$, $g \to R_g^{\lambda}$ is equivalent with a unitary representation if and only if it is already unitary. This happens if and only if there exists an element p of the Weyl group, with $p^2 = 1$, and $p(\lambda) = \lambda'$ (where λ' is the complex contragredient).*

The significance of this result can be understood in view of the following remarks. For $R(s_1) \cdots = R(s_r) = 0$ the $R_g^{\lambda(m,s)}$ are unitarily equivalent with the principal series—but far from identical. It is the former that have the desired analytic continuation, and not the latter. For example, the former are already unitary whenever possible—and this is not the case for the latter. In particular, the former gives us automatically the "complementary series" in unitary form. This is partly due to the fact that the $R_g{}^\lambda$ have all the symmetry of the Weyl group already built in, as indicated by conclusion (5). The property of polynomial growth in (3) is very important for the applications to harmonic analysis of the group.

It is time to reveal for which groups G this result is actually proved. We considered first the case $G = SL(2, R)$ (the case $SL(2, C)$ is completely analogous). The next stage was the consideration of $SL(n, C)$, and here the group $SL(3, C)$ of rank 2 already represents the basic difficulty of $SL(n, C)$.

A further stage already accomplished (but not yet published) is for the other families of complex classical groups $Sp(n, C)$ and $SO(n, C)$. The final stage, which is at present only a goal, would be the consideration of the general semi-simple group G. What remains in the way of this goal? The listener will probably better understand this after the fourth lecture, but at the present suffice it to say that the passage from the complex groups already considered to their real analogues would not seem to be as difficult as the consideration of a seemingly intractable complex group—the exceptional group G_2.

Why do the considerations become successively more complicated and difficult? This seems to be not so much a matter of a weakness of technique as the fact that the situation becomes intrinsically more difficult. This point can be understood as follows. The problems are reducible, to a considerable extent, to questions on the analysis of the nilpotent subgroup N.

In the case of $SL(2, R)$ (or $SL(2, C)$) the group N is Abelian. In the case $SL(n, C)$, this Abelian character is already lost. In the case $Sp(n, C)$ for example, new difficulties are introduced due to the position of N in G etc. We hope these rather vague remarks will become clearer to the listener in the course of these lectures.

Thus we see a curious phenomenon. We are not surprised that the group $SL(2, C)$ does not reveal the general features with respect to a problem of representations theory in the context of—let us say— complex semi-simple groups. We are surprised however when the groups $SL(n, C)$ do not reveal the general features of the problem.

Now to some of the applications of our theorem.

THEOREM 2. *Suppose G is a group of the type described above for which Theorem 1 holds. Let $1 \leqslant p < 2$, and let $k \in L^p(G)$. Then the operator $f \to f*k$ (group convolution) is a bounded operator in $L^2(G)$.*

This seems to be a significant fact in harmonic analysis on the group which is *atypical* of non-compact groups. In particular the analogue of this statement fails whenever G is Abelian and non-compact. It shows that from this point of view the harmonic analysis of G behaves as if G were compact—or nearly so.

It can be readily conjectured that Theorem 1 holds for any real semi-simple group G, and that Theorem 2 holds also for such groups under the restriction that G have finite center.

A final application of our result is as follows. In their book Gelfand and Neumark give a far-reaching construction of irreducible unitary representations of the complex (classical) groups. They claim there that their list exhausts all the irreducible unitary representations of these groups. Purported proofs of this completeness statement have been given by Gelfand and Neumark, Neumark, Berezin, and others. All of these are in error. In fact the family of representations given in Theorem 1 provide us with new irreducible unitary representations for $SO(n, C)$ and $Sp(n, C)$, for appropriate values of s. The detailed case of $Sp(2, C)$ will be considered in Lecture 5.

4. Let us give a brief indication of what is involved in the construction of the uniformly bounded, analytic family of representations. Since the $R^{\lambda(m,s)}$ are unitarily equivalent to the principal series $T^{\lambda(m,s)}$ (when $R(s) = 0$), there exists unitary operators $W(\lambda) = W(m, s)$ for $R(s) = 0$ so that

$$R_g^{\lambda(m, s)} = W(m, s) \, T_g^{\lambda(m, s)}(W(m, s))^{-1}.$$

The problem then is the construction of the $W(\lambda)$. Now because of the symmetry property (5) in Theorem 1 we have

(4.1) $$W^{-1}(p(\lambda)) \, W(\lambda) = A(p, \lambda),$$

where $A(p, \lambda)$ is the intertwining operator for T_g^λ, i.e.,

$$A(p, \lambda) \, T_g^\lambda = T_g^{p\lambda} A(p, \lambda).$$

So the problem of constructing the $W(\lambda)$ involves the question of studying the intertwining operators $A(p, \lambda)$. These operators seem to have a genuine interest on their own—for example they represent natural generalizations of the basic singular integral operators and potential operators to the present context.

The different operators $A(p, \lambda)$ for different p do not commute, but satisfy the "cocycle" relation

(4.2) $$A(pq, \lambda) = A(p, q\lambda)\, A(q, \lambda)$$

wherever p and q are element of the Weyl group. Thus the relations of the Weyl group imply similar relations for the intertwining operators. The sought-for identity (4.1) may be thought of as a coboundary expression for the cocycle—but these formal hints by themselves give little indication of the real problem. The program for analytic continuation can be carried out in the following stages

(1) The construction of the intertwining operator $A(p, \lambda)$.

(2) The study of their relations.

(3) The construction of W in terms of the $A(p, \lambda)$.

(4) The analytic continuation of

$$W(\lambda)\, T_g{}^\lambda (W(\lambda))^{-1}.$$

We shall consider these problems first in the context of $SL(2, C)$—admittedly a simpler case, but one which already reveals at least two essential features of the problem.

Lecture II. $SL(2, C)$

1. Let G be the group of 2×2 complex unimodular matrices. Then every element of g can be represented by

$$g = \begin{pmatrix} a & c \\ b & d \end{pmatrix}, \quad \text{with} \quad ad - bc = 1, \quad a, b, c, d \text{ complex.}$$

The non-trivial irreducible unitary representations of G have been completely classified. These consist of two sets. The *principal series* described more generally in Lecture I and a *complementary series*.

In this case the groups K, A, N can be chosen to be the unitary subgroup, the subgroup of positive diagonal matrices, and the strictly upper triangular subgroup. M is then the diagonal unitary subgroup and

$$MA = \left\{ \begin{pmatrix} \delta_1 & 0 \\ 0 & \delta_2 \end{pmatrix} \right\}, \qquad \text{where} \quad \delta_1 \cdot \delta_2 = 1$$

with $\delta_1 \delta_2$ *complex*. The general unitary representation λ (restricted to MA) is then

$$\lambda(\delta) = |\,\delta_1\,|^{s_1}\,|\,\delta_2\,|^{s_2} \left(\frac{\delta_1}{|\,\delta_1\,|} \right)^{m_1} \left(\frac{\delta_2}{|\,\delta_2\,|} \right)^{m_2}, \qquad \lambda = \lambda(m, s),$$

where $s_1 + s_2 = 0$, $(R(s_1) = R(s_2) = 0)$, and m_1, m_2 are integral with either $m_1 + m_2 = 0$, or $m_1 + m_2 = 1$.

In this case (as in the general cases discussed below) the induced representation $T_g{}^\lambda$ can be realized explicitly as a multiplier representation. That is

(1.1)
$$(T_g{}^\lambda) f(z) = |\,cz + d\,|^{-2+s_2-s_1} \left(\frac{cz + d}{|\,cz + d\,|} \right)^{m_2-m_1} f\left(\frac{az + b}{cz + d} \right),$$

$$\lambda = \lambda(m_1 s).$$

Here the operators $T_g{}^\lambda$ act on the Hilbert space $L^2(dz)$ of square integrable functions on the complex plane, with respect to the additive measure.

The representation (1.1) (with the proviso $R(s_1) = -R(s_2) = 0$) are the *principal series*.

The other series, the *complementary series*, have a definition which is similar—at least on the formal level. We consider the representations

(1.2)
$$f(z) \to |\,cz + d\,|^{-2+2\sigma} f\left(\frac{az + b}{cz + d} \right)$$

defined for $0 < \sigma < 1$, where the inner product of the Hilbert space is now more intricate (and depends for example on σ), namely

(1.3)
$$(f, g)_\sigma = \iint \frac{f(z)\,\overline{g(w)}}{|\,z - w\,|^{2-2\sigma}}\, dz\, dw.$$

It can indeed be shown—and this is not without interest—that the representations (1.2) are unitary with respect to the inner product (1.3).

2. If we are to follow the program outlined at the end of Lecture I our first problem should be to determine the intertwining operators for $g \to T_g{}^\lambda$. In the present case the Weyl group consists of two elements,

and the non-trivial element (as a representative in M' of the coset of M) can be taken to be

$$p = \begin{pmatrix} 0 & +1 \\ -1 & 0 \end{pmatrix}.$$

Its induced action is

$$\begin{pmatrix} \delta_1 & 0 \\ 0 & \delta_2 \end{pmatrix} \rightarrow \begin{pmatrix} \delta_2 & 0 \\ 0 & \delta_1 \end{pmatrix}$$

so that if λ corresponds to $(m_1, m_2; s_1 s_2))$ $p\lambda$ corresponds to (m_2, m_1, s_2, s_1).

So we want to determine the operators $A(p, \lambda)$ with the property that

$$(2.1) \qquad A(p, \lambda) T_g^{\ \lambda} = T_g^{p\lambda} A(p, \lambda), \qquad \text{all} \quad g \in G.$$

Fortunately the problem is tractable because: first, its solution is essentially unique; second, the solution can be determined by requiring (2.1) for a subgroup of G only, for which G takes a particularly simple form.

In fact let $\tilde{B}$ be the subgroup of matrices of the form

$$(2.2) \qquad \begin{pmatrix} a & 0 \\ b & d \end{pmatrix}.$$

This subgroup is the product of two further subgroups

$$(2.3) \qquad \tilde{N} = \left\{ \begin{pmatrix} 1 & 0 \\ b & 1 \end{pmatrix} \right\} \quad \text{and} \quad MA = \left\{ \begin{pmatrix} a & 0 \\ 0 & d \end{pmatrix} \right\}.$$

The action of the representation (1.1) corresponding to elements of $\tilde{N}$ is simply translation

$$f(z) \rightarrow f(z + b),$$

while the action corresponding to elements of the diagonal group MA is (up to a constant multiple) dilation

$$f(z) \rightarrow |d|^{-2+s_2-s_1} \left(\frac{d}{|d|} \right)^{m_1-m_1} f(ad^{-1}z).$$

We are therefore led to the following problem (equivalent with the intertwining identity (2.1)) for g's that are of the form (2.2)).

PROBLEM. *Let s, m be given, with $R(s) = 0$, and m integral. Determine a unitary operator $A(s, m)$ on $L^2(dz)$ which satisfies the following two conditions*

(1) *$A(s, m)$ commutes with translations.*

(2) *With respect to dilation $T_\delta : f(z) \to f(\delta^{-1}z)$, $(\delta \neq 0)$, $A(s, m)$ transforms according to the following rule:*

$$(\tau_\delta)^{-1} A(s, m) \tau_\delta = |\delta|^s \left(\frac{\delta}{|\delta|} \right)^m A(s, m).$$

This problem has a solution, which is unique up to a constant multiple. If $\hat{f}(z)$ denotes the real two-dimensional Fourier transform on $L^2(dz)$, then we can take

$$(A(s, m)f)^\wedge (z) = |z|^{-s} \left(\frac{z}{|z|} \right)^m \hat{f}(z). \tag{2.4}$$

In fact the first condition is equivalent to the fact that $A(s, m)$ can be realized as a multiplier transformation on the Fourier transform side, i.e.,

$$(A(s, m)f)^\wedge (z) = \mu(z) \hat{f}(z).$$

The second condition implies that

$$\mu(\delta z) = |\delta|^{-s} \left(\frac{\delta}{|\delta|} \right)^m \mu(z), \qquad \delta \neq 0;$$

and so,

$$\mu(z) = \text{const} \, x \, |z|^{-s} \left(\frac{z}{|z|} \right)^m.$$

This expression (2.4) for $A(s, m)$ is not appropriate for its further study. We shall need another expression—which can be derived from it by Fourier analysis—viz.,

$$A(s, m)f(z) = \frac{1}{\gamma(s, m)} \int |w|^{-2+s} \left(\frac{w}{|w|} \right)^m f(z - w) \, dw, \tag{2.5}$$

where

$$\gamma(s, m) = i^{|m|} \pi 2^s \frac{\Gamma \left(\frac{|m| + s}{2} \right)}{\Gamma \left(\frac{|m| + 2 - s}{2} \right)}.$$

We are interested in the case $\mathrm{Re}(s) = 0$. For these values of s the integral (2.5) does not converge absolutely but it may be defined as follows. Let f be continuous and of compact support. Then if $\mathrm{Re}(s) = 0$

$$(2.6) \quad A(s, m) f(z) = \lim_{\substack{R(s') > 0 \\ s' \to s}} \frac{1}{\gamma(s', m)} \int |w|^{-2+s'} \left(\frac{w}{|w|}\right)^m f(z - w)\, dw.$$

The $A(p, \lambda)$ may now be determined by setting

$$(2.7) \qquad\qquad A(p, \lambda) = A(s_1 - s_2, m_1 - m_2).$$

These considerations show that the operators $A(p, \lambda)$ just determined satisfy the intertwining property where g is restricted to the triangular subgroup $\tilde{B}$. To prove the intertwining property for all of G it suffices to consider one other element of G which together with $\tilde{B}$ generates G. This element is

$$p = \begin{pmatrix} 0 & 1 \\ -1 & 0 \end{pmatrix}.$$

Now

$$T_p^\lambda f(z) = |z|^{-2+s_2-s_1} \left(\frac{(z)}{|z|}\right)^{m_2-m_1} f(-1/z).$$

The intertwining relation for this element also takes an interesting form. Define $B(s, m)$ to be the operator

$$(2.6) \qquad\qquad B(s, m) f(z) = |z|^{-s} \left(\frac{z}{|z|}\right)^{-m} f(z)$$

and let $\mathscr{T}$ correspond to the inversion

$$(2.8) \qquad\qquad (\mathscr{T}f)(z) = |z|^{-2} f(-1/z), \qquad \mathscr{T}^2 = I.$$

The intertwining identity then becomes

$$(2.9) \qquad\qquad \mathscr{T} A(s, m) \mathscr{T} = B(s, m) A(s, m) B(s, m),$$

which can be verified directly by a change of variables from the integral form of $A(s, m)$ given in (2.6).

These considerations prove that $A(p, \lambda)$ given by (2.7) is indeed the sought for intertwining operator.

3. The second stage in the program outlined at the end of Lecture I is the study of the relations of the intertwining operators. In the present

case matters are trivial. The form (2.4) shows us that the family $A(p, \lambda)$ is additive, i.e.,

$$(2.10) \qquad A(s_1, m_1) A(s_1, m_1) = A(s_1 + s_2, m_1 + m_2).$$

The third stage of the program is the construction of the operator $W(\lambda)$. In this case again the result is immediate. In fact if we set

$$(2.11) \qquad W(\lambda) = A(s_1, m_1)$$

we have

$$W^{-1}(p\lambda) \, W(\lambda) = A(-s_2, -m_2) \, A(s_1, m_1)$$
$$= A(s_1 - s_2, m_1 - m_2)$$
$$= A(p, \lambda),$$

as desired.

The final stage is analytic continuation. What is done here can be reduced to the following lemma.

Let $C(s, m) = B(s, m) \, A(s, m)$

LEMMA. (1) *The family of operators* $\{C(s, m)\}$, *as* s *ranges over imaginary values and* m *over integers is commutative.*

(2) *The function* $s \to C(-s, m_1) \, C(s, m_2)$ *has for each* m_1 *and* m_2 *an analytic continuation in the strip* $-1 < R(s) < 1$, *and*

$$\| C(-s, m_1) \, C(s, m_2) \| \leqslant A(1 + | t | + | m_1 | + | m_2 |)^{|\sigma|},$$

where $s = \sigma + it$.

Strictly speaking the first part of the lemma will not be used here. There are several approaches to the proof of this lemma see [1], [2], and [6]. We notice first that $C(s, m)$ commutes with the dilation $f(z) \to f(\delta z)$, $\delta \neq 0$. Since the dilations form a maximal commutative family on $L^2(dz)$ it follows that the $C(s, m)$ mutually commute. This observation also gives the best, although least direct approach to conclusion (2). We pass to $L^2(dz/| z |^2)$ which is the L^2 space of the multiplicative group of the complex numbers, and for any $\varphi \in L^2(dz/| z |^2)$, we consider its multiplicative Fourier transform ("Mellin transform") given by

$$(M\varphi)(\rho, k) = \int | z |^{-\rho} \left(\frac{z}{| z |} \right)^{-k} \varphi(z) \, \frac{dz}{| z |^2},$$

where ρ is purely imaginary and k is integral. The transformation $C(s, m)$ can then be realized as a multiplier transform, with respect to the Mellin transform, and this multiplier is, up to a fixed factor of absolute value one

$$(2.12) \quad \frac{\Gamma\left(\frac{|k + m| + 1 - (s + \rho)}{2}\right)}{\Gamma\left(\frac{|k + m| + 1 + s + \rho}{2}\right)}.$$

From this the conclusion (2) can be read off from (2.12) by the use of known properties of the gamma function.

We now can give the analytic continuation of the representation

$$R_g^\lambda = W(\lambda)\, T_g^\lambda (W(\lambda))^{-1}.$$

We notice first that R_g^λ is essentially independent of λ for g restricted to the subgroup $\tilde{B}$, (2.2). More particularly $R_g^\lambda = R_g^{\lambda^1}$ if $g \in \tilde{B}$, where $\lambda^1 = \lambda(1, 0, 0, 0)$, or $\lambda^1 = (0, 0, 0, 0)$, when $\lambda = (m_1, m_2, s_1, s_2)$, according to whether $m_1 + m_2$ is odd or even. Thus the analytic continuation for R_g^λ, $g \in \tilde{B}$ is trivial. Now since

$$G = \tilde{B} \cup \tilde{B}p\tilde{B} \quad \left(p \text{ is } \begin{pmatrix} 0 & 1 \\ -1 & 0 \end{pmatrix}\right),$$

it suffices to consider R_p^λ.

But $T_p^\lambda = B(s_1 - s_2, m_1 - m_2)\mathscr{T}$ (see (1.1), (2.6), and (2.7)).

Also $W(\lambda) = A(s_2, m_2)$; we get because of (2.9) (and $\mathscr{T}B(s, m)\mathscr{T} = B(-s, m)$)

$$R_p^\lambda = A(s_1, m_1)\, B(s_1 - s_2, m_1 - m_2)\, \mathscr{T} A(-s_1, -m_1)$$
$$= \mathscr{T} C(s_1, m_1)\, C(-s_1, m_2)\, A(0, m_1 - m_2).$$

Now since $s_1 + s_2 = 0$, and $\mathscr{T}$ and $A(0, m_1 - m_2)$ are fixed unitary operators, the analytic continuation is then a consequence of the lemma.

4. We are now in a position to summarize the results obtained concerning the analytic family of representations for $SL(2, C)$:

(1) R_g^λ is analytic for $\lambda = (m_1, m_2, s_1, s_2)$ with $s_1 + s_2 = 0$, and $-1 < R(s_1) < 1$.

(2) $\sup_g \| R_g^\lambda \| \leqslant A(1 + |t_1| + |m_1|)^{|\sigma_1|}$, where $s_1 = \sigma_1 + it_1$ $(m_1 + m_2 = 0,$ or $1)$.

(3) Since as is easily seen $(R^{\lambda'}_{g-1})^* = R_g^\lambda$ where

$$\lambda' = (m_1, m_2, -\bar{s}_1, -\bar{s}_2),$$

R_g^λ is unitary when either $\lambda = (m_1, m_2, it, it_2)$ (the principal series) or $\lambda = (0, 0, \sigma, -\sigma)$, $-1 < \sigma < 1$ (the complementary series).

Now let $f \in L^1(G) \cap L^2(G)$, and let $F(\lambda) = \int_G R_g^\lambda f(g)\, dg$ denote its group Fourier transform. Then

(4.1) $$\|F(\lambda)\| \leqslant A(1 + |t_1| + m_1|)^{|\sigma_1|}\|f\|_1$$

by conclusion (2).

Also by the Plancheral formula for the group

(4.2) $$\sum_{m_1} \int \|F(\lambda)\|_2^2\, (m_1^2 + t_1^2)\, dt_1 = C\|f\|_2^2.$$

Thus by a known convexity argument

(4.3) $$\sup_{\lambda=(m_1, m_2, it_1, it_2)} \|F(\lambda)\| \leqslant A_p\|f\|_p, \qquad 1 \leqslant p < 2,$$

which shows that the group Fourier transform of an L^p function is uniformly bounded. Thus again by the Plancherel formula for the group, the mapping

$$h \to f * h$$

is bounded on L^2, for any $f \in L^p(G)$, $1 \leqslant p < 2$.

We do not intend to repeat the rather complicated argument by which (4.3) is deduced from (4.1) and (4.2).

Suffice it to say that the explicit form of the Plancherel measure appearing in (4.2) is needed.

More particularly, the fact that it is sufficiently "large" at infinity to compensate for the growth of the bound $A(1 + |t_1| + |m_1|)^{|\sigma_1|}$ appearing in (4.2) is essential.

LECTURE III. THE INTERTWINING OPERATORS

We now turn to the case of the more general groups considered in Lecture I. We shall begin by considering the first stage of our program (outlined in Section 4, Lecture I)—the construction of the intertwining operators. As far as the more formal aspects of this problem are concerned it is best to proceed in greatest generality, that of an arbitrary real semi-simple group (with finite center).

Let G be such a group and g be its Lie algebra. Let $g = k + p$ be a Cartan decomposition where k corresponds to a maximal compact group K, and p is the orthogonal complement of k with respect to the Killing form; a denotes maximal Abelian subspace of p.

For an appropriate positive definite inner product on g, the family of linear transformations $\{ad(a)\}$ is self-adjoint (and of course commutative). This leads to a decomposition of g into the simultaneous (real) eigenspaces of $ad(a)$. Let α be such non-zero simultaneous eigenmap (called a root), then $[H, X] = \alpha(H)X$, $H \in a$, for X in the subspace called g_λ. The roots can be ordered. Choosing one such ordering we let

$$n = \sum_{\alpha > 0} g_\alpha, \qquad \tilde{n} = v = \sum_{\alpha < 0} g_\alpha.$$

With A, N the subgroups corresponding to a and n, respectively, we have the well-known Iwasawa decomposition $G = KAN$.

Now we let M and M' denote, respectively, the centralizer and normalizer of A in K. Then M'/M acts in a natural way on MA and its representations; it is the Weyl group in our context. More particularly let λ be any representation of MA, and let $p \in M'/M$. Define $p\lambda$, by $p\lambda(ma) = \lambda(p^{-1}mpp^{-1}ap)$. This action clearly depends only on the residue class of p in M' modulo M, and so gives the action of the Weyl group M' on the representation of MA. We shall write $B = MAN$, and list the following useful facts about it: B is closed, and N as well as AN are closed normal subgroups of B; every finite dimensional unitary irreducible representation of B is of the form

(1.1) $$man \to \lambda(ma),$$

where $\lambda(ma) = \lambda_M(m)\lambda_A(a)$, and both λ_M, and λ_A are irreducible and unitary (thus λ_A is one-dimensional).

We shall need some further facts about the relation of the subgroup B with respect to G, namely, the Bruhat double-coset decomposition, which can be written as

(1.2) $$G = \bigcup_{p \in M'/M} BpB, \qquad \text{disjoint union.}$$

The elements of a given double coset BpB are not uniquely determined in the form BpB; so we shall look at the situation more closely. The

Weyl group also acts on the roots α, so that $p\alpha$ may be determined by the fact that

$$A\ d(p)(g_\alpha) \subset g_{p\alpha}.$$

Now write

$$n_p = \sum_{\substack{\alpha>0 \\ p^{-1}\alpha<0}} g_\alpha, \qquad n_p' = \sum_{\substack{\alpha>0 \\ p^{-1}>0}} g_\alpha$$

and let N_p and N_p' denote, respectively, the subgroups of N corresponding to n_p and n_p'. Then it can be shown that $N = N_p \cdot N_p'$ and $N_p \cap N_p' = \{\text{identity}\}$, and we have

$$G = \bigcup_{p\in M'/M} N_p \cdot p \cdot B.$$

N_p and N_p' are closed and simply connected subgroups of N. There exists a particular $p_0 \in M'/M$, so that

$$N_{p_0} = N, \qquad \text{and then} \qquad \tilde{N} = V = p_0^{-1}Np_0.$$

2. We come now to the description of the principal series of representations for G.

Let λ be a finite dimensional unitary irreducible representation of $B = MAN$. As stated before

$$\lambda(man) = \lambda_M(m)\,\lambda_A(a),$$

where both λ_M and λ_A are irreducible and unitary. We wish now to describe the induced representation $g \to T_g^\lambda$ of G.

We let db denote a right invariant (Haar) measure of B. Since B is not unimodular we consider the modular function $\mu(b)$, so that $\mu^{-1}(b)\,db$ is a left invariant measure.

We now consider the class of functions f on G with the transformation property[3]

(2.1) $$f(bx) = \mu^{1/2}(b)\,\lambda(b)f(x), \qquad x \in G, \quad b \in B.$$

We norm this space as follows. Let φ be a fixed non-negative continuous function with compact support and with the property that

(2.2) $$\int_B \varphi(bx)\,db = 1, \qquad \text{all} \quad x \in G.[4]$$

---

[3] f takes its values in the finite dimension Hilbert space which is the representation space of λ. We denote the norm in this space by $|\cdot|$.

[4] To prove the existence of such a function we consider first the case $x \in K$. Under these conditions such a φ clearly exists by a simple integration argument. Since $G = BK$, the fact that (2.2) is satisfied for $x \in K$, implies it for all x.

We set

(2.3)
$$\|f\|^2 = \int_G \varphi(x)\,|f(x)|^2\,dx.$$

What is the meaning of this norm? The functions f satisfying (2.1) are completely determined by their values on K. We claim that

(2.4)
$$\|f\|^2 = \int_K |f(k)|^2\,dk.$$

In fact, by a well-known integration formula, for any positive ψ, since $G = BK$,

$$\int_G \psi(x)\,dx = \int_{B \times K} \psi(bk)\,\mu^{-1}(b)\,db\,dk.$$

Using this and the transformation law (2.1), we immediately get (2.4). Now define $T_a{}^\lambda$, $a \in G$, by

(2.5)
$$(T_a{}^\lambda f)(x) = f(xa).$$

Now $\int_B \varphi(bxa)\,db = 1$, all $x \in G$, $a \in G$, and so a repetition of the argument shows that

$$\|T_a{}^\lambda f\| = \|f\|,$$

i.e., $a \to T_a{}^\lambda$ is unitary, $a \in G$.

The representation $a \to T_a{}^\lambda$ just described is a unitary representation on Hilbert space H_λ, whose description we can restate as follows. Start with an "arbitrary" function on K (with values in the representation space of λ) which is in $L^2(K)$ and which transforms as (2.1) for those b that are in K, i.e.,

$$f(mk) = \lambda(m)f(k), \qquad m \in M.$$

Then requiring (2.1) allows one to extend this function (uniquely) to all of G. The norm on H_λ is given by

$$\left(\int_K |f(k)|^2\,dk\right)^{1/2}.$$

We shall refer to this realization of the induced representation (as functions on K) as the "compact picture." It is very useful for many problems such as studying the reduction of $g \to T_g{}^\lambda$ to K, studying traces, etc. We shall need, however, another description, "the non-compact picture," which is analogous to the passage from the circumference of the unit disk to the boundary of the upper-half plane.

Recall the nilpotent subgroup $V = \tilde{N} = p^{-1}Np_0$. Its Lie algebra is given by $v = \sum_{\alpha < 0} g_\alpha$.

Now the double coset Bp_0B is the only one whose dimension equals that of G. Thus by what has already been said Np_0B differs from G by a set of zero measure; the same is therefore true of $p^{-1}Np_0B = VB$, and hence BV differs from G by a set of zero measure. We therefore have

$$(2.6) \qquad\qquad G = BV \qquad \text{almost everywhere}$$

in the sense that almost every element of G can be written (uniquely) in the form $g = bv$, $b \in B$, $v \in V$. Because of (2.6) we have the integration formula

$$\int_G \psi(x) \, dx = \int_{B \times V} \psi(bv) \, \mu^{-1}(b) \, db \, dv$$

(V is unimodular)
and so a similar argument shows

$$(2.7) \qquad\qquad \|f\|^2 = \int_V |f(v)|^2 \, dv.$$

Thus the representation $a \to T_a{}^\lambda$ has a unitarily equivalent realization where H_λ is replaced by the isomorphic Hilbert space of all functions on V for which (2.7) is finite. By the transformation law

$$f(bx) = \mu^{1/2}(b) \, \lambda(b) f(x)$$

and the fact that $G = BV$, this allows one to extend the "arbitrary" function on V to one on G.

It is the second picture, the non-compact one, that will be indispensable for our purposes.

Note that here the representation is

$$\boxed{T_x{}^\lambda f(v) = \mu^{1/2}(vx) \, \lambda(vx) f(v_1)} \quad , \qquad x \in G, \qquad v \in V$$

where v_1 is the unique element of $V \in vx = bv_1$, $b \in B$.

3. We can now give the expression for the intertwining operator $A(p, \lambda)$. In this section our considerations will be purely formal (but not without interest); justifications will be described later.

Let f be a function satisfying the transformation law

$$f(bx) = \mu^{1/2}(b)\,\lambda(b)\,f(x), \qquad x \in G, \qquad b \in B,$$

and define

(3.1) $$F(x) = \int_{n \in N_p} f(p^{-1}nx)\,dn = (A(p, \lambda)f)(x).$$

(This integral, as a matter of fact, almost never converges absolutely!)

Now since the integration in (3.1) is on the left of x (it is in effect a left-convolution in f) and since the action of $T_a{}^\lambda$ is on the right (see (2.5)), it is clear that $A(p, \lambda)\,T_a{}^\lambda = T_a^{p\lambda}A(p, \lambda)$.

What needs to be shown is that, under the assumption that

$$f(bx) = \mu^{1/2}(b)\,\lambda(b)\,f(x),$$

then

$$F(bx) = \mu^{1/2}(b)\,p\lambda(b)\,f(x),$$

where $(p\lambda)(ma) = \lambda(p^{-1}map)$.

Since $B = MAN$, we shall consider M, A, and N separately. First let $m \in M$. Then

$$\int_{N_p} f(p^{-1}nmx)\,dn = \int_{N_p} f(p^{-1}mn'x)\,dn,$$

where $n' = m^{-1}nm$. Since we have already pointed out that each $m \in M$ normalizes N_p, and since M is compact, the module[5] of the transformation $n \to m^{-1}nm$, must be 1. So

$$\int_{N_p} f(p^{-1}nmx)\,dn = \int_{N_p} f(p^{-1}mnx)\,dn;$$

however $p^{-1}m = p^{-1}mpp^{-1}$, and $p^{-1}mp \in M$, since M' also normalizes M. Thus $f(p^{-1}mu) = \lambda(p^{-1}mp)\,f(p^{-1}u)$ by the transformation law (2.1) and we see that

(3.2) $$F(mx) = \lambda(p^{-1}mp)\,F(x).$$

[5] That is Jacobian determinant of transformation.

The argument for $a \in A$ is similar except now the module of $n \to a^{-1}na$ is no longer one; in fact if $\mu_p(a)$ is defined by

$$\int_{N_p} g(n) \, dn = \mu_p(a) \int_{N_p} g(ana^{-1}) \, dn, \qquad a \in A,$$

one can verify that $\mu(a) = \mu(p^{-1}ap)(\mu_p(a))^2$, and this follows from the identity

$$\mu(\exp H) = \exp\left(\sum_{a>0} c_\alpha \alpha(H)\right),$$

where $c_\alpha = \dim g_\alpha$.

Thus an argument, silimar to the one used for M, shows that

$$(3.3) \qquad\qquad F(ax) = \mu^{1/2}(a) \, \lambda(p^{-1}ap) \, F(x), \qquad x \in A.$$

It remains to consider an n_0 in N and to show that

$$(3.4) \qquad\qquad F(n_0 x) = F(x), \qquad n_0 \in N.$$

Now write $n_0 = n_1 \cdot n_2$, where $n_1 \in N_p$, $n_2 \in N_p'$. The consideration of n_1 is a trivial change of variables in the integral (3.1), and finally, when $n_0 = n_2$, we have

$$f(p^{-1}nn_2x) = f(p^{-1}n_3n'x) = f(p^{-1}n_3 pp^{-1}n'x) = f(p^{-1}n'x).$$

Here

$$n_3 \in N'_p, \, n' \in N_p.$$

The first equality is valid since we have $N = N_p N_p' = N_p' N_p$, and the last equality because

$$(\mu^{1/2}\lambda)(p^{-1}n_3 p) = 1, \qquad \text{since} \quad p^{-1}N_p'p \subset N.$$

Finally the mapping of $N_p : n \to n'$, given by

$$nn_2 = n_3 n', \qquad n_2, \qquad n_3 \in N_p',$$

has module 1, since all the groups N, N_p, N_p' are unimodular, as a standard integration argument shows. This proves (3.4).

We should like now to transform the intertwining integral (3.1) into a form appropriate to the study of functions on V, as in the definition of the "non-compact picture" described above. For this purpose note that the fact that $G = BV$ almost everywhere indicates that $p^{-1}n = b_1z$, where $n \in N_p, b_1 \in B$, and $z \in V$. As n runs over N_p, z will describe a certain set O_p, in V. Since $f(b_1zx) = \mu^{1/2}(b_1) \lambda(b_1) f(zx)$, we can write the integral (3.1) as

$$(3.3) \qquad \int_{O_p} K(z, \lambda) f(zx)\, dz$$

for a suitable kernel $K(z, \lambda)$ and measure dz on O_p. The advantage of (3.3) is that it shows that the intertwining operator is a convolution on the nilpotent group V, and so we see that the study of the intertwining operators is connected with the Fourier analysis of V.

In certain important special cases the formula (3.3) takes a simple and explicit form. We have in mind the case that occurs when p is a *basic reflection*, that is, when the subgroup is given in terms of the Weyl reflection by a root α, and this reflection changes the sign of only those roots which are multiples of α.

In this case it can be shown that essentially $O_p = V_p$ where $V_p = p^{-1}N_p p$.

V_p is the subgroup of V whose lie algebra v_p is given by

$$v_p = \sum_{\substack{\beta > 0 \\ p^{-1}\beta < 0}} g_{-\beta}\,.$$

Then the integral takes the form

$$(3.4) \qquad \int_{V_p} \mu^{1/2}(vp)\, \lambda^{-1}(vp) f(vx)\, dv,$$

where $\mu^{1/2}(vp)\, \lambda^{-1}(vp) = \mu^{1/2}(b_1)\, \lambda^{-1}(b_1)$ with $vp = b_1v_1$, and $b_1 \in B$, $v_1 \in V$; dv is Haar measure on the group V_p.

Before completing all these formal considerations it might be well to give an explicit example. Let $G = SL(3, \mathbb{C})$, and let λ be trivial on M. Then λ may be identified with a triple of complex numbers s_1, s_2, s_3, where $R(s_1) = R(s_2) = R(s_3) = 0$, and $s_1 + s_2 + s_3 = 0$.

There are two basic reflections, p_1 and p_2, whose induced action is $p_1(s_1, s_2, s_3) = (s_2, s_1, s_3)$ and $p_2(s_1, s_2, s_3) = (s_1, s_3, s_2)$. The element p_0 that changes the signs of all the roots has the induced action $p_0(s_1, s_2, s_3) = (s_3, s_2, s_1)$. We can choose for V the subgroup of lower triangular unipotent matrices

$$v = \begin{bmatrix} 1 & 0 & 0 \\ v_1 & 1 & 0 \\ v_2 & v_3 & 1 \end{bmatrix} = (v_1, v_2, v_3).$$

Then

$$(3.5) \qquad A(p_1, \lambda) f(x) = \int |v_1|^{-2+s_1-s_2} f(x_1 + v_1, x_2, x_3) \, dv_1 ,$$

$$(3.5') \qquad A(p_2, \lambda) f(x) = \int |v_3|^{-2+s_2-s_3} f(x_1, x_2 + v_3 x_1, x_3 + v_3) \, dv_3 ,$$

and

$$(3.5'') \qquad A(p_0, \lambda) f(x) = \int |v_1 v_3 - v_2|^{-2+s_1-s_2} |v_2|^{-2+s_2-s_3}$$

$$\times f(x_1 + v_1, x_2 + v_2 + v_3 x_1, x_3 + v_3) \, dv_1 \, dv_2 \, dv_3 .$$

4. We pass to more rigorous considerations and now limit ourselves to the case when G is a *complex* semi-simple group, and p is still a basic reflection. Then $v_p + n_p + [v_p, n_p]$ is a three-dimensional complex Lie subalgebra of g which is isomorphic to the Lie algebra of $SL(2, \mathrm{C})$. Write

$$V = V_p \cdot V_p',$$

which is the analogue of the decomposition $N = N_p N_p'$ already used. We may then write symbolically $L^2(V) = L^2(V_p) \otimes L^2(V_p')$. Observe also that, if $x \in V$, the action of the integral (3.4) is only on the V_p projection of x. It can then be seen that, as far as $x \in V_p$, the integral (3.4) is identical with its analogue for $SL(2, \mathrm{C})$. In that case the intertwining integral has already been rigorously constructed, by dividing first by an appropriate factor $\gamma(s, n)$, and then by a passage to the limit, approaching the case of unitary characters λ by non-unitary ones. (See (2.6) in Lecture II.) Therefore these considerations apply also in the case of a basic reflection for complex semi-simple G.

We intend to study the case of the intertwining operators corresponding to the general elements of the Weyl group by using the fact that the basic reflections already generate the Weyl group.

LECTURE IV. THE REFLECTIONS

1. The situation which has been obtained for any complex semi-simple group G can now be described as follows. Among the positive

roots we can choose a basis of positive *simple* roots. Let α_1, α_2 ,..., α_r be these roots. Then the reflection

$$p_j = p_{\alpha_j}(\beta) = \beta - 2\alpha_j \frac{(\beta, \alpha_j)}{(\alpha_j, \alpha_j)}$$

is then a basic reflection.

Now according to the theory of the Weyl group, $W = M'/M$, W is generated by p_1, p_2 ,..., p_r; the group is also characterized by the following relations among those generators

(1.1) $p_j{}^2 = 1,$

(1.2) $(p_i p_j)^{n(i,j)} = 1,$

where $n(i, j)$ is an integer which depends only on the angle between α_i, α_j .

We have

(a) $n(i, j) = 2,$ if α_i and α_j are orthogonal.

(g) $n(i, j) = 3,$ if angle is $2\pi/3$.

(c) $n(i, j) = 4,$ if angle is $3\pi/4$.

(d) $n(i, j) = 6,$ if angle is $5\pi/6$.

Now the $A(p_j, \lambda)$ have already been constructed by a reduction of the problem to the case of $SL(2, C)$. We need to study now the relations among those $A(p_j, \lambda)$ which in effect are a consequence of the relations among the elements of the Weyl group and the desired fact that

(1.3) $A(pq, \lambda) = A(p, q\lambda) A(q, \lambda).$

Let us therefore fix p_i, and p_j. Let g' be the subalgebra of g generated over C by $g_{\alpha_i} g_{\alpha_j}$, $g_{-\alpha_i}$, $g_{-\alpha_j}$. Then g' is a complex semi-simple algebra of rank 2 which is isomorphic to:

    in case (a), $sl(2, c) \oplus sl(2, c)$,
    in case (b), $sl(3, c)$,
    in case (c), $sp(2, c)$,
    in case (d),   type G_2 .

Let G' be the subgroup of G corresponding to the lie subalgebra g'. Since the subgroups V_{p_1}, V_{p_2} can be seen to be exactly those whose Lie algebras are $g_{-\alpha_1}$ and $g_{-\alpha_2}$, respectively, an examination of the formula (3.4) (in Lecture III) has the following consequence.

All problems of the relations concerning $A(p_i, \lambda)$ and $A(p_j, \lambda)$ can be reduced to the corresponding problems for the groups of rank 2. Fortunately, in the complex classical groups, the only subgroups of rank 2 that arise correspond to the first three cases, and exclude the case of G_2.

It is for this reason, and in order to simplify the exposition, that we shall limit most of our remaining considerations to an explicit description of what happens for the groups $SL(3, C)$ and $Sp(2, C)$. (An understanding of what happens in G_2 also would certainly solve the problem for all complex semi-simple groups.

We shall make another simplification in much of the written formulae by assuming that the part of λ coming from M is trivial. (It is in any case a character of a compact Abelian group, and no analytic continuation with respect to it will be performed. This restriction is made however only for notational convenience.)

2. We now consider the case of $SL(3, C)$.

The two intertwining operators $A(p_1, \lambda)$ and $A(p_2, \lambda)$ corresponding to the two simple roots have already been written (except for the normalizing factor $1/\gamma$) in formulae (3.5) and (3.5'). The result to prove, which corresponds to $(p_1 p_2)^3 = 1$, is

$$(2.1) \qquad A(p_1, p_2 p_1 p_2 \lambda)\, A(p_2, p_1 p_2, \lambda)\, A(p_1, p_2 \lambda)$$
$$= A(p_2, p_1 p_2 p_1 \lambda)\, A(p_1, p_2 p_1 \lambda)\, A(p_2, p_1 \lambda).$$

If we write

$$A_1(s_1 - s_2) = A(p_1, \lambda),$$

and

$$A_2(s_2 - s_3) = A(p_2, \lambda),$$

then after some simplification the relation becomes

$$(2.2) \qquad\qquad \{A_1(s)\, A_2(s)\}_s \qquad \text{is a commutative family.}$$

Now a closer study of the Fourier analysis of the group V, in this case the three-dimensional group of strictly upper triangular matrices in $SL(3, C)$, shows that the statement (2.2) is equivalent with the statement

$$(2.2') \qquad\qquad\qquad \{B(s)\, A(s)\}_s$$

is a commutative family when $A(s) = A(s, 0)$, $B(s) = B(s, 0)$ in the notation of Section 2, Lecture II, dealing with $SL(2, C)$.

Thus the present statement (2.2′) is an immediate consequence of the lemma proved there.

3. The study for $Sp(2, C)$ is similar to a certain extent, but is more complicated because the corresponding nilpotent group V is one further step removed from the Abelian case then in the case $SL(3, C)$.

In the case $Sp(2, C)$ we can arrange matters so that MA is the group

$$\begin{pmatrix} \delta_1^{-1} & & & \\ & \delta_2^{-1} & & 0 \\ \hline & & \delta_1 & \\ 0 & & & \delta_2 \end{pmatrix}, \quad \text{with} \quad \delta_1 \neq 0, \quad \delta_2 \neq 0.$$

Assuming as before that λ is trivial on M, then

$$\lambda = |\delta_1|^{s_1} |\delta_2|^{s_2} ,..., R(s_1) = R(s_2) = 0.$$

We can let $V =$ the subgroup

$$(3.1) \qquad \begin{pmatrix} 1 & -v_1 & & \\ & 1 & & 0 \\ \hline & & 1 & \\ 0 & & v_1 & 1 \end{pmatrix} \begin{pmatrix} I & & 0 \\ & & \\ \hline v_2 & v_3 & \\ v_3 & v_4 & I \end{pmatrix} = (v_1 , v_2 , v_3 , v_4).$$

The subalgebras $g_{-\alpha_1}$ and $g_{-\alpha_2}$ are, respectively, those which generate the (complex) one-parameter groups $\{(v_1 , 0, 0, 0)\}$ and $\{(0, v_2 , 0, 0)\}$.

The action of

$$p_1 \quad \text{corresponds to} \quad p_1(s_1 , s_2) = (s_2 , s_1)$$

and

$$p_2 \quad \text{corresponds to} \quad p_2(s_1 , s_2) = (-s_1 , s_2).$$

After some reduction we can write

$$A(p_1 , \lambda) = A_1(s_2 - s_1), \qquad A(p_2 , \lambda) = A_2(s_1).$$

The relation corresponding to $(p_1 p_2)^4 =$ identity then becomes

$$A_2(s_1) A_1(s_1 + s_2) A_2(s_2) A_1(s_2 - s_1) = A_1(s_1 - s_2) A_2(s_2) A_1(s_1 + s_2) A_2(s_1).$$

This is equivalent with the statement that

(3.2) $\{A_1(s)\,A_2(s)\,A_1(s)\}_s$ is a commutative family.

Again we must appeal to the Fourier analysis of the group V. We can then transform this statement into an equivalent one

(3.2′) $\{A(s)\,Q(s)\,A(s)\}_s$ is a commutative family.

Now everything is taken in $L^2(dz)$. $A(s)$ is the operator arising in $SL(2, C)$, and $Q(s)$ is the multiplication operator

$$Q(s)f(z) = |\,1 - z^2\,|^{-s} f(z).$$

To study the situation we use the theory for $SL(2, C)$. Let

$$a = \frac{1}{\sqrt{2i}} \begin{pmatrix} 1 & 1 \\ 1 & -1 \end{pmatrix},$$

and define

$$T_a f(z) = \frac{1}{2} f\left(\frac{z + 1}{z - 1}\right) |\,z - 1\,|^{-2}.$$

(This unitary operator arises, of course, from the representation of the principal series of $SL(2, C)$ corresponding to the trivial λ, $\lambda \equiv 1$.) A simple calculation shows that

(3.3) $T_a Q(s)(T_a)^{-1} f(z) = 4^{-s} |\,z\,|^{-s} |\,z - 1\,|^{2s} f(z),$

while the intertwining property of $A(s)$ also shows that

(3.4) $T_a A(s)(T_a)^{-1} = M(s)\,A(s)\,M(s),$

where $M(s)f = |\,z - 1\,|^{-s}f(z)$. The last identity is closely analogous to the identity (2.9) in Lecture II. Thus altogether the problem (3.2′) is reduced to the statement that

(3.5) $\{M(s)\,A(s)\,B(s)\,A(s)\,M(s)\}_s$ is a commutative family.

This last fact can be reduced to the facts that $\{M(s)\,A(s)\}_s$ and $\{A(s)\,B(s)\}_s$ are commutative families as well as the fact that B and M commute. (See Lecture V, Section 2.)

In this very sketchy way we have summarized the ideas behind the proof of the relations satisfied by the intertwining operators (at least for the complex classical groups).

4. We come now to the problem of analytic continuation. We shall in this lecture give the results for $SL(n, C)$, making use of the same notational convenience used before, namely, disregarding that part of λ which comes from the compact M (in this case an $n - 1$ torus). The group MA can be chosen to be the diagonal subgroup

$$\begin{pmatrix} \delta_1 & & & \\ & \delta_2 & & \\ & & \ddots & \\ & & & \delta_n \end{pmatrix}, \qquad \text{with} \quad \delta_1 \cdot \delta_2 \cdots \delta_n = 1.$$

The λ are $|\delta_1|^{s_1} \cdots |\delta_n|^{s_n}$, with $s_1 + s_2 + \cdots + s_n = 0$. There are $n - 1$ basic reflection $p_1, p_2, \ldots, p_{n-1}$, which can be realized so that p_j is essentially the permutation matrix which permutes j, with $j + 1$, leaving the other letters $1, 2, \ldots, j - 1, j + 2, \ldots, n$ fixed. Then

$$p_j(s_1, s_2, \ldots, s_n) = (s_1, \ldots, s_{j-1}, s_{j+1}, s_j, s_{j+2}, \ldots, s_n).$$

We can write $A(p_j, \lambda) = A_j(s_j - s_{j+1})$. With this notation fixed then we can write

$$(3.6) \qquad W(\lambda) = A_{n-1}(s_1) A_{n-2}(s_1) \cdots A_1(s_1) A_{n-1}(s_2) \cdots A_2(s_2) \cdots A_{n-1}(s_{n-1})$$

(a product of $n(n - 1)/2$ operators).

It can be verified that

$$(3.7) \qquad W(p(\lambda))^{-1} W(\lambda) = A(p, \lambda)$$

and to do this it suffices to verify (3.7) for $p = p_j$.

Finally let $R_x{}^\lambda$ be defined by

$$(3.8) \qquad R_x{}^\lambda = W(\lambda) T_x{}^\lambda W^{-1}(\lambda).$$

This "normalization" of the principal series has a remarkable property, which indeed characterizes it.

Let G_0 be the subgroup of G whose entries in the last column (except for the diagonal entry) are all zero. Then

$$(3.9) \qquad R^\lambda |_{G_0} \qquad \text{is independent of} \quad \lambda.$$

So in order to carry out the analytic continuation in λ it suffices to carry it out for the fixed element p_{n-1}, since $G = G_0 \cup G_0 p_{n-1} G_0$.

However if s_1 , s_2 ,..., s_{n-2} are fixed and s_{n-1} , s_{n-2} variable

$$(s_1 + s_2 \cdots + s_n = 0),$$

then the analytic continuation of $W(\lambda) \, T^\lambda_{p_{n-1}} \, W^{-1}(\lambda)$ is easily reducible to the corresponding continuation already carried out in $SL(2, C)$. Since $R^\lambda = R^{p\lambda}$, the situation is completely symmetric in all the variables s_1 , s_2 ,..., s_n , and thus we can continue separately in any pair of variables, and hence altogether. The reader is referred to the literature for further details.

LECTURE V.   THE OTHER COMPLEX CLASSICAL GROUPS

1. The analytic continuation in the case of $SL(n, C)$ is a completely unique and natural construction because the operators $W(\lambda)$ in that case are characterized by transforming the principal series into a "normalized one" which has maximal constancy in λ, in particular, this constancy is on the subgroup G_0 where all the representations are already irreducible!

The existence of this subgroup G_0 is a special property of $SL(n, C)$, and is not shared by the other groups. The analytic continuation for the other groups, where this has been done, can then be carried out only by partial analogy. For these groups one is now faced by the additional problem of constructing some "square roots" of families of operators. This is done in the following lemma which is basic for our purposes. Let $\mathscr{H}$ be a Hilbert space.

LEMMA 1.   *Suppose $s \to \Phi(s)$ is a continuous mapping from $\mathrm{Re}(s) = 0$ to unitary operators on $\mathscr{H}$. Suppose also*

(1) $\Phi(-s) = (\Phi(s))^{-1}$.

(2) $\{\Phi(s)\}_s$ *is a commutative family.*

(3) $\Phi(s)$ *has a relative analytic continuation in the strip $-1 < R(s) < 1$ in the sense that if it_1 is fixed $s \to \Phi(s) \, \Phi(-s + it_1)$ is analytic in $-1 < R(s) < 1$ and at most of polynomial growth in $| \mathrm{Im}(s)| + | t_1 |$.*

Then there exists a family $\Psi(s)$, satisfying the same properties as Φ in the strip, $-1 < R(s) < 1$, so that

(a) $\Psi^2(s) = \Phi(s)$.

(b) *If $\overline{X}$ commutes with Φ then $\overline{X}$ commutes with Ψ.*

To prove the lemma we consider first the special case when Φ itself has a continuation (as bounded operators) in the strip $-1 < R(s) < 1$. (This is not the case in the main application below.)

Set

$$L(s) = \int_0^s \Phi'(\xi)\, \Phi(-\xi)\, d\xi = \int_0^s \Phi'(\xi)(\Phi(\xi))^{-1}\, d\xi.$$

We can verify that $e^{L(s)} = \Phi(s)$. Set

$$\Psi(s) = e^{L(s)/2}.$$

This solves the problem in that case. Notice that Ψ is actually given the whole strip, $-1 < R(s) < 1$.

Next let

$$K(s_1, s_2) = \Phi(s_1)\, \Phi(-s_2)\, \Phi(-s_1 + s_2).$$

Then according to our assumptions for each fixed s_1, $\mathrm{Re}(s_1) = 0$ $K(s_1, s_2)$ has an analytic continuation in the strip $-1 < R(s_2) < 1$. Similarly for each fixed s_2, $R(s_2) = 0$, $K(s_1, s_2)$ has a continuation in the strip $-1 < R(s_1) < 1$. Then by a known argument $K(s_1, s_2)$ has a joint continuation in the tube

$$|R(s_1)| + |R(s_2)| < 1.$$

Now let $s_1 = -s_2 = s$ in K. Then we have that $K(s) = \Phi^2(s)\, \Phi(-2s)$ has a continuation in the strip $-\frac{1}{2} < R(s) < \frac{1}{2}$. Let $K_{1/2}(s)$ be the function guaranteed by the special case of the lemma already proved with the property that $(K_{1/2}(s))^2 = K(s)$.

Set

$$\Psi(s) = \Phi(s/2)\, K_{1/2}(-s).$$

This Ψ can be easily seen to satisfy the conclusions of the lemma.

2. We shall now apply the lemma to the family

$$\Phi(s) = A_1(s)\, A_2(s)\, A_1(s).$$

See (3.2) in Lecture IV.

We saw previously that this family could be successively transformed to $A(s)\, Q(s)\, A(s)$ and then to

(2.1) $M(s)\, A(s)\, B(s)\, A(s)\, M(s).$

Now clearly the family (2.1) satisfies the conditions (1) of the lemma; (2), that it is a commutative family was already established. Thus we need to verify the "relative" analytic continuability of (2.1).

Now a simple variant of the lemma of analytic continuation of Lecture II (see Section 2) shows that $B(s) A(s)$ has a relative analytic continuation in the sense required, i.e., for each fixed t_1

$$s \rightarrow B(s) A(s) B(-s + it_1) A(-s + it_1)$$

is analytically continuable in the strip $-1 < R(s) < 1$, and of polynomial growth in s and t_1. The same is then true for $M(s) A(s)$ instead of $B(s) A(s)$. (In fact $M(s) A(s) = \tau B(s) A(s) \tau^{-1}$ where $(\tau f)(v_1) = f(v_1 - 1)$.) It is also true for $A(s) M(s)$ instead of $M(s) A(s)$, as a simple argument involving adjoints will show. However, if $R(s_1 + s_2) = 0$, we have

$$M(s_1) A(s_1) B(s_1) A(s_1) M(s_1) M(s_2) A(s_2) B(s_2) A(s_2) M(s_2) =$$

$$M(s_1) A(s_1) B(s_1) M(s_2) A(s_2) A(s_1) M(s_1) B(s_2) A(s_2) M(s_2) =$$

$$M(s_1) A(s_1) M(s_2) B(s_1) A(s_1) A(s_2) B(s_2) M(s_1) A(s_2) M(s_2) =$$

$$M(s_1) A(s_1) M(s_2) A(s_2) B(s_2) B(s_1) A(s_1) M(s_1) A(s_2) M(s_2).$$

The first step arises by commuting the factors $A(s_1) M(s_1)$ with $M(s_2) A(s_2)$; the second step is the result of commuting $B(s_1)$ with $M(s_2)$, $B(s_2)$ with $M(s_1)$, and $A(s_1)$ with $A(s_2)$; the last step because we commute the factors $B(s_1) A(s_1)$ with $A(s_2) B(s_2)$.

The final form is the required form, because each of its factors,

$$M(s_1) A(s_1) M(s_2) A(s_2), \quad B(s_2 + s_1), \quad \text{and} \quad A(s_1) M(s_1) A(s_2) M(s_2),$$

has the desired analytic continuation.

This shows that we can find the appropriate square root of $A_1(s) A_2(s) A_1(s)$. We call it Ψ.

We are now in a position to write the $W(\lambda)$ operator for $Sp(2, C)$. We set

$$(2.2) \qquad W(\lambda) = \Psi(s_1) \Psi(s_2) A_1(-s_1).$$

Let us first check that

$$(2.3) \qquad W^{-1}(p\lambda) W(\lambda) = A(p, \lambda).$$

It suffices to verify this for the basic reflections p_1 and p_2. (Recall the form of these given in Section 3 of Lecture IV.) Since $\Psi(s_1)$ commutes with $\Psi(s_2)$, the factor $\Psi(s_1)\,\Psi(s_2)$ is unchanged by p_1. Thus

$$W(\lambda) = W(p_1\lambda)\,A_1(s_2 - s_1)$$

and (2.3) is proved for p_1.

However

$$
\begin{aligned}
W^{-1}(p_2\lambda)\,W(\lambda) &= A_1(-s_1)(\Psi(-s_1))^{-1}\,\Psi(s_1)\,A_1(-s_1) \\
&= A_1(-s_1)\,\Psi^2(s_1)\,A_1(-s_1) \\
&= A_1(-s_1)\,A_1(s_1)\,A_2(s_2)\,A_1(s_1)\,A_1(-s_1) \\
&= A_2(s_1) = A(p_2, \lambda);
\end{aligned}
$$

therefore (2.3) is completely proved.

We now let G_0 be the subgroup generated by V, MA (MA is a Cartan subgroup) and the reflection p_2.

We claim next that

$$(2.4)\qquad R_x{}^\lambda = W(\lambda)\,T_x{}^\lambda W^{-1}(\lambda), \qquad \text{is independent of } \lambda \text{ for } x \in G_0.$$

The proof of this is similar to the proof of the analogous fact for $SL(n, C)$ except for one additional point.

Consider the element $p_1 p_2 p_1$ of the Weyl group. Since

$$A(p_1 p_2 p_1, \lambda) = A(p_1, p_2 p_1\lambda)\,A(p_2, p_1\lambda)\,A(p_1, \lambda)$$

we have

$$A(p_1 p_2 p_1\lambda) = A_1(s_1 + s_2)\,A_2(s_2)\,A_1(s_2 - s_1).$$

However it can be seen that $T_{p_2}^\lambda$ depends only on s_1. Thus we write $T_{p_2}^\lambda = T_{p_2}^{s_1}$. So we have

$$A(p_1 p_2 p_1, \lambda)\,T_{p_2}^{s_1} = T_{p_2}^{s_1}A(p_1 p_2 p_1, \lambda),$$

and setting $s_1 = 0$ we have

$$A_1(s)\,A_2(s)\,A_1(s)\,T_{p_2}^{id} = T_{p_2}^{id}A_1(s)\,A_2(s)\,A_1(s),$$

that is, $A_1(s)\,A_2(s)\,A_1(s)$ commutes with $T_{p_2}^{id}$ and so, by conclusion (b) of the lemma, the square root $\Psi(s)$ also commutes with $T_{p_2}^{id}$.

Once (2.4) is proved, the analytic continuation of R^λ is then reduced to that of $R^\lambda_{p_1}$, since

$$G = G_0 \cup G_0 p_1 G_0 .$$

However

$$R^\lambda_{p_1} = \Psi(s_1)\, \Psi(s_2)\, A(-s_1)\, T^\lambda_{p_1} A(s_1)\, \Psi(-s_1)\, \Psi(-s_2).$$

Let $s_j = \sigma_j + it_j$, and assume that $\sigma_1 + \sigma_2 = 0$. Then the factors $\Psi(s_1)\, \Psi(s_2)$ have bounded continuations (as long as $-1 < \sigma_1 < 1$), by the lemma; similarly for the factors $\Psi(-s_1)\, \Psi(-s_2)$; finally

$$A(-s_1)\, T^\lambda_{p_1} A(s_1)$$

has a continuation by the argument for $SL(2, C)$.

Since $R^\lambda = R^{q\lambda}$, for any q in the Weyl group, the same holds also for the line $\sigma_1 - \sigma_2 = 0$, $-1 < \sigma_1 < 1$. We therefore obtain the continuation in the tube whose basis is the square

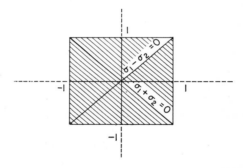

which is the convex hull of the indicated diagonals. To the basic fact that

(2.5) $R^\lambda = R^{q\lambda},$ q any element of the Weyl group,

must be added the much simpler but still decisive observation that

(2.6) $(R^\lambda_{x^{-1}})^* = R^{\lambda'}_x ,$

where $*$ denotes the Hermitian conjugate of the operator, and, if $\lambda = (s_1, s_2)$, then $\lambda' = (-\bar{s}_1, -\bar{s}_2)$ (with $-$ denoting the complex conjugate).

In fact the identity (2.6), when $R(s_1) = R(s_2) = 0$, is nothing but a rephrasing of the statement that, for those λ, R^λ is unitary (as it is by its construction). From that special case we obtain the more general form of

(2.6) by analytic continuation, since both sides of (2.6) are analytic functions of $\bar{s}_1$ and $\bar{s}_2$.

3. An immediate consequence of (2.5) and (2.6) is the following. Suppose $s_1 = \sigma_1 + it_1$, $s_2 = \sigma_2 + it_2$ lies in the tube whose basis is pictured above. Then the representation corresponding to (s_1, s_2) is unitary if there exists an element q of the Weyl group so that

$$(3.1) \qquad q(s_1, s_2) = (-\bar{s}_1, -\bar{s}_2).$$

We list in Table 1 the resulting possibilities. In order to test (3.1), it suffices to consider only the elements q, so that $q^2 = $ identity. There are six elements of this form (the Weyl group in question is of order 8).

TABLE 1

	Element q of Weyl group	Its action	Parameterization of unitary representations
1	$q = $ identity	$(s_1, s_2) \to (s_1, s_2)$	(it_1, it_2)
2	$q = p_1$	$(s_1, s_2) \to (s_2, s_1)$	$(\sigma + it, -\sigma + it)$
3	$q = p_2$	$(s_1, s_2) \to (-s_1, s_2)$	(σ, it)
4	$q = p_1 p_2 p_1$	$(s_1, s_2) \to (s_1, -s_2)$	(it, σ)
5	$q = p_2 p_1 p_2$	$(s_1 s_2) \to (-s_2, -s_1)$	$(\sigma + it, \sigma - it)$
6	$q = (p_1 p_2)^2$	$(s_1, s_2) \to (-s_1, -s_2)$	(σ_1, σ_2)

We shall comment about these collections of representations of $Sp(2, C)$ in order:

(1) This case corresponds of course to the "principal series" constructed by Gelfand and Neumark.

(2) This corresponds to a "complementary series" constructed by Gelfand and Neumark.

(3) Another complementary series. It however *does not* appear in the list of Gelfand and Neumark. It could have been constructed by the same technique used in (2) and as such does not reveal anything essentially new; it, however, already demonstrates the incompleteness of their list.

(4) Equivalent with those in (3).

(5) Equivalent with those in (2).

(6) Here are the essentially new representations whose construction seems to require the main parts of the ideas detailed above.

The characters of these representations can be obtained in a known manner from the formulas for the principal series, by analytic continuation. From this one can see that the cases (1), (2), (3), and (6) are inequivalent. It also follows from the general theory of Bruhat that, in general, these representations are irreducible. Thus, in case (6), the representation corresponding to (σ_1, σ_2) is irreducible if $\sigma_1 \neq 0$, $\sigma_2 \neq 0$, $\sigma_1 \neq \sigma_2$, and $\sigma_1 \neq -\sigma_2$.

LECTURE VI (not presented). SOME ADDITIONAL REMARKS

1. The construction given in the previous lecture for $Sp(2, C)$ is typical of a more general construction which can be given for $Sp(n, C)$.

For the other complex classical groups an analogous procedure works. We shall limit ourselves to some remarks for the groups $SO(2n + 1, C)$.

The construction given for $Sp(2, C)$ had at its starting point the commutative family

$$(1.1) \qquad\qquad A_1(s)\, A_2(s)\, A_1(s).$$

There is, however, another commutative family, namely

$$(1.2) \qquad\qquad A_2(s)\, A_1(2s)\, A_2(s).$$

(That (1.2) is a commutative family can easily be seen to be equivalent with the statement that (1.1) is a commutative family.)

Now if we start with the family $A_2(s)\, A_1(2s)\, A_2(s)$, then find its square root, and proceed analogously as before we obtain another analytic continuation. This will be the construction typical for $SO(2n + 1, C)$, and this can be understood by the fact that locally

$$Sp(2, C) \sim SO(5, C).$$

2. The consideration of analytic continuation for the principal series of the complex classical groups allows one to prove the L^p convolution theorem (Theorem 2, Lecture I) for these groups, in a manner very similar to the way it was done in the case of $SL(2, R)$.

The problem that is raised is whether this theorem is valid for *any* semi-simple group (with finite center).

An indication that this might be true is given by the following theorem valid for such general groups

THEOREM. *Suppose $g(x) \in L^p(G)$, $1 \leqslant p < 2$, and suppose that g is bi-invariant, i.e., $g(x) = g(k_1 x k_2)$, $k_1, k_2 \in K$. Then the operator $f \to g * f$ is bounded on $L^2(G)$.*

A proof is as follows:

Let $\varphi_0(x)$ be the *positive* spherical function given by

$$(2.1) \qquad \varphi_0(x) = \int_K e^{-\rho H(xk)} \, dk$$

Then according to estimates of Harish-Chandra it follows that

$$\| \varphi_0(x) \|_q < \infty, \qquad \text{where} \qquad q > 2, \quad 1/p + 1/q = 1.$$

Then because of the identity

$$(2.2) \qquad \int_K \varphi_0(xky) \, dk = \varphi_0(x) \, \varphi_0(y)$$

it is an easy exercise to verify that

$$(2.3) \quad g * \varphi_0 = c\varphi_0, \quad \text{where} \quad c = \int_G g(y) \, \varphi_0(y^{-1}) \, dy, \quad |c| \leqslant \| g \|_p \| \varphi \|_q,$$

since g is bi-invariant.

Let $T(f) = g * f$, and let M denote the multiplication operator $(Mf)(x) = \varphi_0(x) f(x)$. It then follows by (2.3) that

$$M^{-1} T M$$

is a bounded operator of $L^\infty(G)$ to itself, i.e.,

$$(2.4) \qquad \| M^{-1} T M f \|_\infty \leqslant A \| f \|_\infty .$$

By a duality argument, using the fact that roughly speaking both M and T are self-dual, we have

$$(2.5) \qquad \| M T M^{-1} f \|_1 \leqslant A \| f \|_1 .$$

A known convexity argument then allows one to interpolate between (2.4) and (2.5), giving

$$\| Tf \|_2 \leqslant A \| f \|_2 ,$$

which is the desired result.

REFERENCES AND BIBLIOGRAPHIC COMMENTS

We make no attempt to give here the vast background literature of the subject; instead we quote only those papers which contain the details of the arguments sketched above.

1. R. A. KUNZE AND E. M. STEIN. Uniformly bounded representations..., *Amer. J. Math.* **82** (1960), 1–62.
2. R. A. KUNZE AND E. M. STEIN, II. *Amer. J. Math.* **83** (1961), 723–786.
3. R. A. KUNZE AND E. M. STEIN. Intertwining operators for the principal series of representations of complex semi-simple groups, preprint. Princeton University, 1964.
4. R. A. KUNZE AND E. M. STEIN, III, *Amer. J. Math.* **89** (1967), 385–442.
5. R. LIPSMAN. Uniformly bounded representations of $SL(2, C)$, *Amer. J. Math.*, to appear.
6. E. M. STEIN. Analysis in matrix spaces and some new representations of $SL(N, C)$. *Ann. of Math.* (November 1967).

LECTURE 2.   The study of $SL(2, C)$ is closely related to that given for $SL(2, R)$ ins[1]. For $SL(n, C)$ see [2]. A recent detailed study of $SL(2, C)$ can be found in [5]. The particular. argument sketched for the lemma in section 2 can be found more generally in the paper [6] That paper deals in part with the analytic continuation for certain *degenerate* series for $SL(n, C)$ analogous to $SL(2, C)$.

LECTURE 3.   For a fuller presentation of the ideas given here see [4]. The formulae (3.5), (3.5′), and (3.5″) as well as another approach to the determination of the intertwining operators are found in the unpublished paper [3].

LECTURE 4.   For further details see [2] and [4].

LECTURE 5 (AND 6).   The results sketched here will appear in detail in a forthcoming paper of Kunze and the author.

Mathematics $2.95

YALE MATHEMATICAL MONOGRAPHS

Yale University Press announces a new series of paperbound monographs in mathematics, chosen by an editorial committee made up of mathematicians from the faculty of Yale University. Each volume treats a topic of unusual interest at the forefront of current mathematical research. The first four volumes, listed below, are based on lectures given by outstanding young mathematicians who were chosen to speak on topics in modern mathematics under the auspices of the annual James K. Whittemore Lectures in Mathematics at Yale.

1. **Euler Products**   Robert P. Langlands
2. **Analytic Continuation of Group Representations**
   Elias M. Stein
3. **Algebraic Spaces**   Michael Artin
4. **Group Theory and Three-Dimensional Manifolds**
   John Stallings

ANALYTIC CONTINUATION OF GROUP REPRESENTATIONS

by Elias M. Stein, professor of mathematics at Princeton University.

Mr. Stein discusses the problem of finding analytic continuation of irreducible unitary representations (belonging to the principal or the complementary series) of a certain class of semi-simple Lie groups G. The problem is not simple because analytic continuation is not unique.

The case when $G = SL(2,C)$ is discussed in detail, while only an outline of an argument is given for the case when G is a more general complex semi-simple Lie group. These results constitute the main part of a series of three papers which Professor Stein has published in the *American Journal of Mathematics* in collaboration with Professor R. A. Kunze. Later results in the theory of analytic continuation and indications of future development are also included.

ISBN 0-300-01428-7

YALE
UNIVERSITY
PRESS
NEW HAVEN
AND LONDON